John Alleyne Macnab

Song of the Passaic

John Alleyne Macnab

Song of the Passaic

ISBN/EAN: 9783337180713

Printed in Europe, USA, Canada, Australia, Japan

Cover: Foto ©Thomas Meinert / pixelio.de

More available books at **www.hansebooks.com**

SONG OF THE PASSAIC.

LIVINGSTON, ESSEX COUNTY.

And who can paint the glist or gleam,
That lurks upon that flowing stream?

Song of the Passaic.

With Illustrations.

Also, a Descriptive Sketch and Map of the River
and its Tributaries.

By John Alleyne Macnab.

" So green is the grass, so clear is the stream,
So mild is the mist and so rich is the beam,
That beauty should never to other lands roam,
But make on the banks of our river its home."

Thomas Davis.

New York.
WALBRIDGE & CO.,
1890.

THE MACKENZIE PRESS,
WALBRIDGE & CO.
17 TO 27 VANDEWATER ST.,
NEW YORK.

Song of the Passaic.

I.

The great Jehovah wisely planned
All things of earth, divinely grand;
And. in His way, all nature tends
To laws divine, to serve His ends.

II.

The rivers run, and none shall know
How long their waters yet may flow;
We read the record of the past,
While time withholds the future cast.

NEW PROVIDENCE, UNION COUNTY.

The rivers run, and none shall know
How long their waters yet may flow.

III.

The buds and leaves change shade and form;
Winds list, and lo! comes calm or storm.
Birds wing their flight from North to South;
The spring floods come, and Summer's drought;
And ev'ry thing, with touch of time,
Will gather mould and frosts of rime.

IV.

Yet, in their flowing to the sea,
The rivers fill their destiny;
And to the measure of their lays,
Run on and on, through endless days;
As old as. hills, and yet as new
As verdured fields, they wander through.

STANLEY, MORRIS COUNTY.

As old as hills, and yet as new
As verdured fields, they wander through.

V.

And in the rise, the light and glow
Of grand old rivers, in their flow
From distant hills, through dales and lea,
The fair Passaic seeks the sea—
An Indian name,[1] which signifies
That, broad and fair and fruitful, lies,
Between the sea and mountain source,
The valley where its waters course.

VI.

And who can paint the glist or gleam
That lurks upon that flowing stream;
Or measure beauty of its glades;
Or weigh the tints of var'ing shades;
Or follow furrowed ways, deep-fraught
With wondrous lines, by water wrought?

1 See Explanatory Notes.

ROSELAND, ESSEX COUNTY

From distant hills, through dales and lea.
The fair Passaic seeks the sea.

VII.

Or trace the forms that come and go,
Where tow'ring trees sway to and fro:
And bushy banks throw on its tides,
The flittings of their fringing sides?

VIII.

Or grasp, from closing shades of night,
The ling'ring rays of fading light;
Or steal, where trails of moonbeams prance,
In midnight whirl of wat'ry dance:
Or hold the light, when stars are dimmed,
And mist with rays of light are rimmed;
Or keep, or fix the lights that dart,
And mingling seem, yet keep apart?

HORSENECK, MORRIS COUNTY.

And bushy banks throw on its tides,
The flittings of their fringing sides.

IX.

And He, who bade the rivers flow,
Gave sparkle to Passaic's glow ;
Made it to run through dale and glen,
Through mead and wood, and leafy fen,
Until its slow and lazy tide
O'erflows its banks, on either side,
And vast expanse of "Waste Lands"[2] fill
With ooze of water, at its will,
And marshes, sodden, low and dank,
O'ergrown with grasses, wild and rank.

GREAT PIECE MEADOWS, MORRIS COUNTY.

And vast expanse of "Waste Lands" fill
With ooze of water, at its will.

X.

Then onward flows, with good intent
To wider channels in descent;
While other tributes [3] seek, and pour
Their wealth of flow unto its store;
And greater range of watershed
Gives broader surface to its spread.

TWO BRIDGES, PASSAIC COUNTY.

While other tributes seek, and pour
Their wealth of flow unto its store.

XI.

And, from its wealth of broad green fields,
A fuller measure drainage yields;
And, by the lave of flowing stream,
The soils with greater bounties teem;
The valleys are more fair with bloom;
The verdure richer with perfume.
And denser is the growth that weaves.
In closer web, the fringe of leaves.

FAIRFIELD, ESSEX COUNTY.

And denser is the growth that weaves,
In closer web, the fringe of leaves.

XII.

Its placid tides now disengage,
When tranquil flow gives way to rage;
And loosened water disenthralls,
And pours its mass down waterfalls.

I.—LITTLE FALLS, PASSAIC COUNTY

Its placid tides now disengage,
When tranquil flow gives way to rage.

XIII.

And up above the whirl and toss
Of maddened water, swings across
A timbered bridge, with shingled roof
And covered sides, thrown high, aloof,
O'er ragged, deep, rock-bound defile;
Nor pier nor 'butments serve to guile
The rushing tides, in hurried flow,
To basins, cleft in rocks below.

XIV.

While hemlock trees, that drape the edge
Of time-sered rocks, o'erhang the ledge
And drop their cones; so light and small,
That one would scarcely note their fall,
Were not a legend [1] still in vogue,
That each cone holds a fairy rogue.

II.—LITTLE FALLS, PASSAIC COUNTY

The rushing tides, in hurried flow,
To basins, cleft in rocks below.

XV.

And overhead, in arch[5] of stone,
There flows a tide, so slow and prone
To will of men, whose genius planned
A waterway; that runs through land,
Where ne'er before had water flowed,
Or boats by horse or mules been towed.

XVI.

And circles of the archway's rim
Cast shad'wy forms, like spectres grim;
And frowning rocks, o'ergrown with moss,
Shine in the moonbeams' silv'ry gloss;
While circling eddies woo and press
The waters into tranquilness.

III.—LITTLE FALLS, PASSAIC COUNTY.

And circles, of the archway's rim,
Cast shad'wy forms, like spectres grim.

And oaken trees, and fir-tipped pine,
And green and rugged cedars, line
The ragged cliff; where winds, astir,
Are rustling oak-leaves 'gainst the fir;
And beetling crags, aglint and steep,
Hang o'er the waters, purling, deep;
And, from the archways' crumbling walls,
The dripping of the leakage falls;
And, in its slumbers, 'mid the sheen
Of flowing rims and wooded screen,
There lies an Isle, where quartz and schist
Hold glittering gems of amethyst;"
And, over all, there lurks a sound—
The distant roar where waters bound.

IV.—LITTLE FALLS, PASSAIC COUNTY.

There lies an isle; where quartz and schist
Hold glitt'ring gems of amethyst.

XVIII

Still, on and on, its tides descend,

In rippling flow and winding trend,

And, in its sweep, it sparkles bright—

A silver ribbon bathed in light.

RYLE PARK, PASSAIC COUNTY.

And, in its sweep. it sparkles bright—
A silver ribbon bathed in light.

XIX.

Then, under dizzy heights of bridge,
And circling round the wooded ridge,
With rolling surface, broader spread,
It passes cities of the dead.

LAUREL GROVE, PASSAIC COUNTY.
Then, under dizzy heights of bridge,
And circling 'round the wooded ridge.

XX.

When, measures of its hoarded store

Recede, and plash on either shore;

And stubborn walls* but hold it back,

That raceways may recall, to rack,

From its walled-up, impeded course,

Industrial share of motive force;

And lead, in angling conduits down,

A power that serves the busy town.

I.—PATERSON, PASSAIC COUNTY.

And stubborn walls but hold it back,
That raceways may recall, to rack—

XXI.

The riven waters rush distract,
And boldly leap the cataract;
Awing, by their majestic might
And thrilling grandeur of their flight,
They plunge sheer down, in deep abysm,
With awful roar and rumbling rythm;
And break, on rocks that intervene,
Into a mass of whitened sheen.

XXII.

And, in the whirl where eddies flow,
The foam-flakes ride, white as the snow;
And clouds of spray and vap'ry mist
Rise up and fall, sun-hued and kissed
By Iris' light, whose magic ray
Throws arches o'er the misted way;
And rainbow colors interlock
In circles 'round the creviced rock.

11.—PASSAIC FALLS, PATERSON.
The riven waters rush distract,
And boldly leap the cataract.

XXIII.

And rigid, grav, deep-walled and grand,
The grim old rocks, like sentries, stand;
While deaf'ning thunder of its bound
Reverberates, and prisoned sound
Rolls back and forth, till sound-wave fills,
And echoes linger on the hills.

XXIV.

Then, through the rocks, in broken rifts,
The waters splurge, in swaying drifts;
While on a crest, rock-rimmed and hilled,
A brown shaft [10] stands; its tablets filled
With names of heroes, battle-slain;
While waters murmur sad refrain
And ripple on, and circuit-lines
The Ysland,[11] clothed with stately pines.

III.—VALLEY OF THE ROCKS, PATERSON.

Then through the rocks, in broken rifts,
The waters splurge, in swaying drifts.

XXV.

Then, to the rythm of its own song
And measured flow, it winds along;
While river, valley, mountain, plain,
Together form one vast domain,
And tree-girt shores reflect their trace,
In shad'wy lines, in its embrace;
And waters broaden, full and free,
And gleam and glow in Lake Dundee;[12]
While brightest beams of early dawn,
Kiss graves and spires of Cedar Lawn.[13]

DUNDEE, BERGEN COUNTY.

And waters broaden, full and free,
And gleam and glow in Lake Dundee.

XXVI.

Then waters course through gate and race

And roll o'er wall,[14] with sheeled face;

And, passing fields of fruit and tillage,

Greet namesake city,[15] hamlet, village;

And, winding on, 'mid ebb and flow

Of ocean tides that come and go,

The famous river yields its sway

Unto the long deep-channeled bay;

And, feeling touch of new emotion,

It gives its tides to those of ocean.

CLIFTON, PASSAIC COUNTY.

Then waters course, through gate and race,
And roll o'er wall, with sheeted face.

XXVII.

Ah, who will paint the scenes that lie,
Where waters of that river ply?
And whose the pen to paint so clear
A picture, that will be as dear
As tides, that run and gleam, and glow
With silv'ry sheen of constant flow.

XXVIII.

And who will speak for shady bowers?
And who will sing of fragrant flowers,
That line its banks, while lilies throw
Their blooms from wat'ry depths below?
And who will sing of brooks or rills,
That run to rivers from the hills?
And what of plumaged birds, that wing
Their flight from shore to shore, and bring
A wealth of song, sung without words;
A song that only comes from birds?

PASSAIC, PASSAIC COUNTY.

Ah ; who will paint the scenes, that lie
Where waters of that River, ply.

XXIX.

And what of Summer, with green vales
Of so rich yield, or fertile dales
By waters laved, and clad in dress,
That Nature gives with lavishness?
And what of Autumn, with leaves turned
From green to russet, sered, and burned
To brown and crimson, gray and gold,
Upon its surface onward rolled?

XXX.

And what of Winter, with its gloss,
Deep fringing banks with icy floss;
Or fields, so lately brown and trite,
All clad in raiment, snowy white;
When Nature sleeps so sound, and brings
Renewed life with its 'wakenings?
And what of all that treasured lore—
The legendary tales of yore?

DELAWANNA, PASSAIC COUNTY.

To brown and crimson, gray and gold,
Upon its surface onward rolled.

XXXI.

And what of busy water-wheels;
The crank of shafts, or hum of reels?
And what of conquered tongues of fire.
Yielding to water its burning pyre?
And what of cooling draughts that spring
From mountain-lakes, refreshening
Alike to man and beast, and bird,
Who drink its waters undisturbed?

XXXII.

Could I but sing of all its ways!
That grand old river! all its lays,
And all loved scenes; but I must yield
To other pens the fruitful field.

IV.—WEST SIDE PARK, PATERSON.

Could I but sing of all its ways!
That grand old River! all its lays—

INDEX TO LAKES AND PONDS SHOWN ON MAP.

Round Pond, Greenwood Lake. Macopin Pond and Green Pond, being named on the map, have no numbers attached.

RAMAPO WATERSHED.

No. on Map	Name.	County.		Elevation.
	Round Pond.	..Orange,	N. Y.	662 feet.
1.	Mt. Basha Lake.	"	"	851 "
2.	Slaughter Pond	"	"	1 054 "
3.	Cranberry Pond	"	"	1,013 "
4.	Echo Lake.	"	"	709 "
5.	Island Pond.	"	"	963 "
6.	Green Pond.	"	"	991 "
7.	Car Pond...	"	"	706 "
8	Tuxedo Lake	"	"	557 "
9.	Little Negro Pond..	... "	"	775 "
10.	Negro Pond, or Lake Po Take	Rockland N. Y. and Passaic, N. J.		610 "
11.	Rotten Pond....	"	"	537 "
12.	Crooked Pond	Bergen,	"	410 "
13.	Franklin Lake.	"	"	417 "
14.	Pompton Lake artific'a	.Passaic,	"	210 "

WANAQUE WATERSHED

No. on Map	Name.	County.		Elevation.
—	Greenwood Lake..	Orange, N. Y. and Passaic, N. J.		621 feet.
15.	Stirling Lake..	..Orange,	N. Y.	749 "
16.	Little Cedar Pond .	"	"	1,029 "
17.	Sheppard's Pond .	Rockland, N.Y. and Passaic, N. J.		637 "
18.	Tice's Pond....	"	"	490 "
19.	Bearfort Pond	"	"	1,238 "
20.	Terrace Pond.	"	"	1,393 "
21.	Mud Pond	"	"	340 "

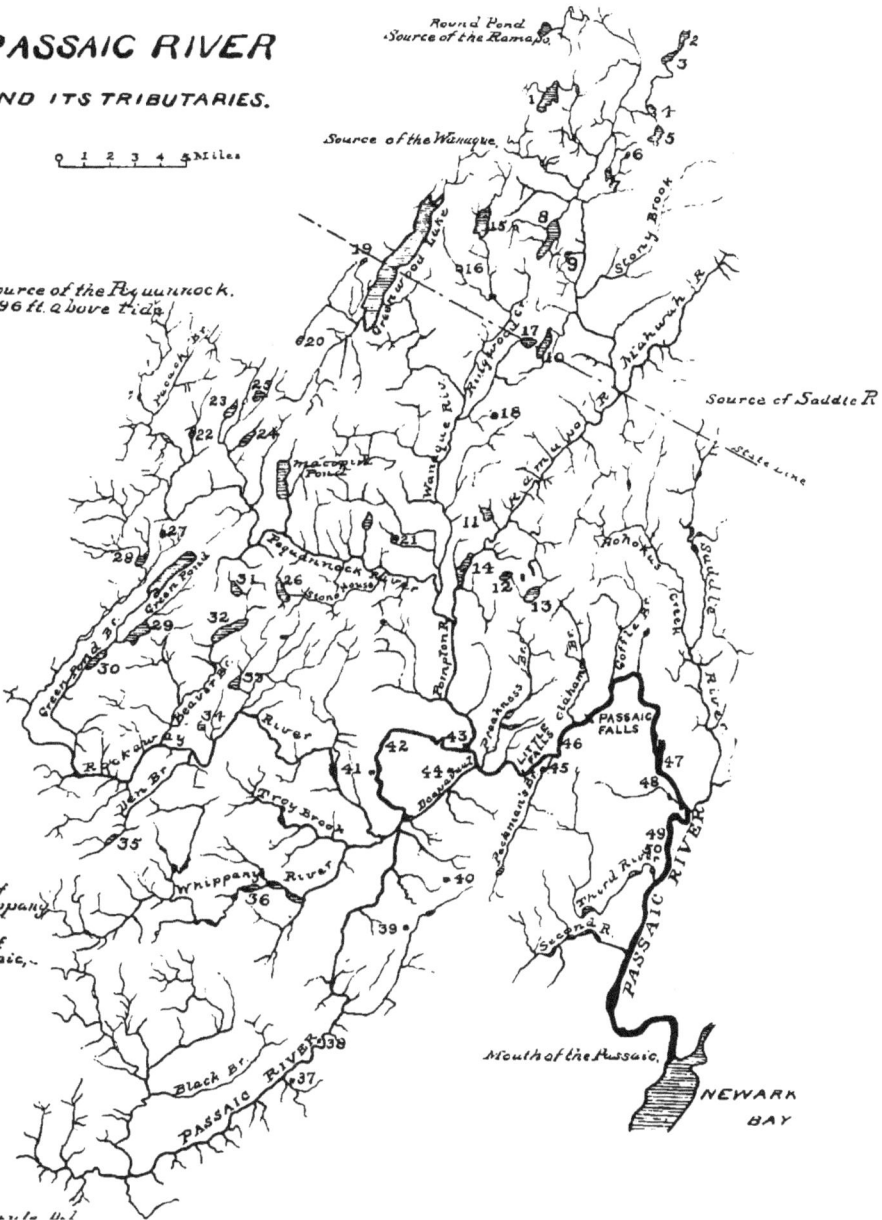

PASSAIC RIVER

AND ITS TRIBUTARIES.

0 1 2 3 4 5 Miles

Round Pond
Source of the Ramapo

Source of the Wanaque

Source of the Rockaway.
1496 ft. above tide.

Source of Saddle R.

State Line

Pompton R.

PASSAIC FALLS

LITTLE

Black Br.

PASSAIC RIVER

Second R.

Third River

Mouth of the Passaic.

NEWARK BAY

Stony Brook

Wanaque River

Ringwood Cr.

Pequanock River

Rockaway River

Beaver Br.

Troy Brook

Whippany River

Green Pond

Greenwood Lake

Macoppin Pond

Wynokie R.

Ramapo R.

Hohokus Creek

Saddle River

Saddle R.

Goffle Br.

Goffle Br.

Deepavack Br.

Peckman's Br.

of Whippany

of Passaic,

PEQUANNOCK WATERSHED

No. on Map.	Name.	County.		Elevation.
22.	Dunker Pond......	..Passaic,	N. J.	1,013 feet.
23.	Buckabear Pond.....	"	"	995 "
24.	Hank's Pond.................	"	"	1,033 "
25.	Cedar Pond..................... ..	"	"	1,110 "
—	Macopin Pond...	"	"	893 "
26.	Stickle Pond.........	..Morris,	"	786 "

ROCKAWAY WATERSHED.

No. on Map.	Name.	County.		Elevation.
27.	Mooseback Pond............... ....Morris,		N. J.	815 feet.
28.	Petersburgh Pond......................	"	"	775 "
—	Green Pond...........................	"	"	1,048 "
29.	Denmark Pond	"	"	818 "
30.	Middleforge Pond...	"	"	708 "
31.	Durham Pond.	"	"	880 "
32.	Split Rock Pond.......................	"	"	815 "
33.	Dixon's Pond.....:....................	"	"	560 "
34.	Duck Pond...........................	"	"	521 "
35.	Shongum Pond.......................	"	"	698 "

WHIPPANY WATERSHED.

No. on Map.	Name.	County.		Elevation.
36.	Speedwell Lake........................Morris,		N. J.	312 feet.

REFERENCE TABLE.

This table shows the location on the river where the views were taken.

Little Falls and Passaic Falls are named on the map; and the four illustrations at the former place, as also the four illustrations at the latter place, have no numbers attached.

The vignette title of this book shows two of the three rocks at Two Bridges, which are referred to in the descriptive sketch of the river.

It will be noticed that some of the numbers on the map, as well as a small star near the number, are placed some distance from the river. This indicates that the town, or village proper, is at some distance from the river; the limits of the town or village, however, extend to the banks of the river.

NOTES, EXPLANATORY OF THE POEM.

1.—Passaic is an Indian word, and in its generally accepted significance means a Valley; and it is so defined by "Webster."

2.—What is here meant by the term, "Waste Lands," is all of that tract of low meadow shown and designated on maps of New Jersey as the "Great Piece Meadows." Locally it is generally called the "Big Piece;" but, properly speaking, the Big Piece lies on the Essex County side of the river, and that portion lying on the Morris County side is called, locally, the "Low Lands." The Great Piece Meadows cover an area of territory five miles square in extent. A goodly portion of it is quite heavily timbered; and, as the elevation of the river at Two Bridges—the commencement of the meadows—is only one foot greater than at the top of Beattie's dam, three miles below, it will be seen that the floods that come from its great area of drainage readily overflow these low lands. This applies with equal force to the low lands further up the river, lying between the Great Piece Meadows and Chatham. The territory embraced by the latter, which is generally called the Hanover Meadows, is fully as large as that of those first named. The excavations now in progress for draining these meadows must be carried to completion before an intelligent judgment can be pronounced as to the success of the work.

3.—The Pompton River unites with the Passaic at Two Bridges, as shown in illustration on the opposite page. The bridge to the left crosses the Passaic River, while the one to the right crosses the Pompton River.

4.—This legend, it is said, applies only to the cones of the hemlock tree. These cones, it will be remembered, are no larger than an ordinary sewing thimble.

5.—The Morris Canal crosses the river at Little Falls in a stone aqueduct, at an elevation of fifty feet above the river. Considering that it was built about half-a-century ago, it is a grand work, not only in its mechanical excellence and architectural beauty, but from the engineering skill displayed in the construction. The freezing, every winter, of its water-soaked stones, has given it a severe trial; and there are many who predict that the time is not far distant when the whole structure will give way. The outer stones of its upper or western rim have already fallen to the river below; and were it not for the strengthening frame, or wooden conduit added to it a few years ago, the probabilities are that the prediction would have been verified ere this. It is constructed of sandstone, which was quarried in the near vicinity. This quarry gained some distinction from the fact that it furnished the stone for Trinity Church on Broadway, New York. The ledge on which this quarry is located has since been opened on the other side of the river.

NOTES.

6.—Numerous and beautiful specimens of amethyst have been found in the quartz and mica-schist, in the Basin or Valley of the Rocks, at Little Falls. (See Catalogue of Minerals, page 4, Part 1, Vol. II; Final Reports of State Geologist; 1889.)

7.—Referring to Laurel Grove and Holy Sepulchre Cemeteries. The former is comparatively a new cemetery, beautifully located on the side of a sloping hill. The latter, and older, is a Catholic Cemetery, and is situated about one-half mile nearer to the city of Paterson than is Laurel Grove.

8.—For reasons that will be apparent, the writer has used the word "wall" in the poem rather than the word "dam." What is meant by the word "wall" is the dam of the "Society for Establishing Useful Manufactures." This Society was incorporated in 1792 The dam diverts the water of the river into raceways, and thus furnishes motive power to many factories. However, this is only a small portion of the power now used; as many of the factories having water power, also have, in addition, steam power ; and the city has grown to such an extent, that a great number of the factories and many of them the largest in the city—are far removed from the river or the raceways

9.—It is of frequent occurrence for rainbows to span the Chasm at Passaic Falls; the shifting rays of the sun throw the curve of the rainbows sometimes longitudinal with the Chasm, and sometimes directly across it. And while it is not frequent, yet it is not rare, to see a rainbow high up in the air reflected from one lower down.

10.—This refers to the Passaic County Soldiers' and Sailors' Monument, erected in 1870. It is of brown stone, and is surmounted by a white marble statue of a Union soldier. The white tablets, on its four sides, contain the names of two hundred and forty-eight Soldiers and Sailors, who sacrificed their lives for their and our country in the late Civil War.

11.—This is the island adjoining the West Street Bridge, at Paterson, long known as Temperance Island, but of late years known as Little Coney Island

12.—Dundee Lake is formed by damming the river. The present dam was erected about the year 1860, to furnish water power for some contemplated mills at the city of Passaic; many mills have since been erected. Previous to the erection of the present dam a smaller one existed, and the adjoining portion of Bergen County was then known as "Slaughter Dam." The banks of the river as they existed before the present dam was built, are readily distinguished by the occasional stumps of trees that protrude above the water.

13.—Cedar Lawn Cemetery is on the Passaic County side of the river, at Lake Dundee. This cemetery overlooks a great portion of Bergen County, and comprises several hundreds of acres. It is beautifully situated and is subject to unremitting care.

NOTES.

14 – Dundee Dam, which has already been spoken of in reference to Dundee Lake, is here referred to.

15. The term "Namesake City," refers to the city of Passaic, and the words "Hamlet" and "Village" in the poem, immediately following "Namesake City," are intended to mean Delawanna and Belleville.

In the selection of the illustrations, the author has tried to choose such as will convey a general idea of the river, throughout its length; and, at the same time, give those views that are accessible to any one who may desire to ride, or walk, along the roadways that follow the course of the river; rather than give illustrations—even though they might surpass some of those selected—that are inaccessible or can be reached only by a row-boat.

NOTE CONCERNING NAMES.

For the purpose of designating one of the illustrations in this book—in the absence of any fixed name – the writer has taken the liberty of calling that portion of the river lying between Lincoln Bridge and the bridge of the D, L. & W. R. R. Co., Laurel Grove. And although he finds on maps of the watershed of the Ramapo River, in Orange County, New York a pond that is known as "Nigger" Pond, yet he has taken the liberty of calling this sheet of water, "Little Negro Pond"—partly to distinguish it from a larger pond that lies in an adjacent county, that bears a similar name, but chiefly because of the better orthography of the name.

In regard to the names of lakes and ponds adopted by the writer. He has chosen those that have been the most generally accepted, although, in some instances, they may differ from those given on maps—even of recent issue— of the Geological Survey. It is to be regretted that some lakes and ponds have two or three names. "Lake Po-Take" and "Negro Pond" refer to the same body of water; the former has the right of priority, although circumstance has more definitely fixed the latter.

It will also be noticed, that some of the elevations of lakes or ponds as herein noted, differ in a trifling measure from those given on maps of the Geological Survey. This is due to the fact that the figures quoted here are from more recent surveys, and indicate a *normal* elevation. Sheppard's Pond, during a dry season, lies wholly in Passaic County, N. J., while in a wet season it expands into Rockland County, N. Y. The map shows it in its expanded condition, while the figures given indicate its normal elevation.

DESCRIPTIVE SKETCH OF THE PASSAIC RIVER.

The following sketch of the river has been prepared partly from personal knowledge, partly from reports of the State Geologist, partly from data kindly furnished by the Civil Engineer and Topographer who made the surveys and maps of the Passaic river and valley—and from other sources. And for the reason that the writer found many who had but little knowledge of the river— its source, its tributaries, its length or elevations—he has endeavored to make this sketch not only accurate, but full.

The Passaic River rises in the Great Swamp near Mendham, Morris County, New Jersey, and retains its name down to Newark Bay. The word Passaic is of Indian origin, and signifies a valley. The total length of the river—from source to mouth—is eighty-one and a quarter miles. The area of drainage at Millington, the outlet of the Great Swamp, is 56.6 square miles; at Chatham Bridge, 25½ miles below its source, it is 99.8 square miles; at Little Falls, 52½ miles below its source, it is 774.2 square miles; at Passaic Falls, at Paterson, 57 miles below its source, it is 796.9 square miles; and at its mouth, at Newark, it is 949.1 square miles. Its elevation above tide water at foot of Dundee Dam, is 6 feet; at top of Dam, 27 feet; foot of Passaic Falls, 40 feet; top of Passaic Falls, 110 feet; foot of Little Falls, 118 feet; top of Beattie's Dam, 158 feet; mouth of Pompton River, at Two Bridges, 159 feet; mouth of Rockaway River, 163 feet; Chatham Bridge, 177 feet; outlet of Great Swamp, at Millington, 221 feet; while its source, at Mendham, is 600 feet above tide. Tide water flows to the city of Passaic, which is 13 miles above its mouth, and the river is navigable to that city.

The area of drainage above Little Falls, is made up from that of its branches, the Ramapo, Wanaque, Pequannock, Rockaway and Whippany Rivers, and the Upper Passaic. Below Little Falls, it receives the drainage of Peckman's Brook, and the Oldham Creek, and at the city of Passaic, that of Saddle River. Many other streams, the larger of which are named on the map, add their quota to the river's total.

The Ramapo River rises in Orange and Rockland Counties, New York. The Wanaque River rises in Orange County, in the same State. The Pequannock River rises in Morris, Passaic, and Sussex Counties, N. J. The Rockaway River has its whole course in Morris County. The Whippany River rises near Morristown, in the same county. The Ramapo, Wanaque and Pequannock Rivers unite at Pompton, and form Pompton River, which runs into the Passaic at Two Bridges. The Ramapo River derives its waters from Round Pond, Mount Basha Lake, Slaughter Pond, Cranberry Pond, Echo Lake, Island Pond, Green Pond, Car Pond, Tuxedo Lake and Little Negro Pond. All of the foregoing lakes or ponds lie in Orange County, New York. It also receives the waters of Negro Pond, or Lake Po Take, which lies partly in Rockland County, N. Y., and partly in Passaic County, N. J.; Rotten Pond, in Passaic County; Crooked Pond and Franklin Lake, in Bergen County, and Pompton Lake (artificial), in Passaic County.

The Wanaque River derives its waters from Greenwood Lake, which lies partly in Orange County, N. J., and partly in Passaic County, N. J.; Stirling Lake and Little Cedar Pond, in Orange County, N. Y.; Sheppard's Pond, which lies partly in Rockland County, N. Y., and partly in Passaic County, N. J.; Tice's Pond, Bearfort Pond, Terrace Pond and Mud Pond, in Passaic County.

The Pequannock River receives its flow from Dunker Pond, Buckabear Pond, Hank's Pond, Cedar Pond and Macopin Pond in Passaic County, and Stickle Pond in Morris County. The newer name of Macopin Pond is Echo Lake, but the older name of Macopin still clings to it, and while it is admitted that the conformation of the valleys that surround this sheet of water are such that an "Echo range" is produced, still it should be borne in mind that echo ranges are of such frequent occurrence that no particular importance should be attached to it. The Indian name of Macopin (pronounced Mawcopin), was given, undoubtedly, to this beautiful sheet of water, not because of its echo range, but because of that far more phenomenal and characteristic feature, its Whisper Alley —or the ease with which one can converse across its waters, not only from side to side, but from end to end. It is a pity —nay, it was a crime upon nomenclature to change its name to the inferior one of Echo Lake. True, the Indians have long since passed away, but—

> " Their names are on your waters,
> And ye cannot wash them out."

The Rockaway River draws its supply from Mooseback Pond, Petersburg Pond, Green Pond, Denmark Pond, Middleforge Pond, Durham Pond, Split Rock Pond, Dixon's Pond, Duck Pond and Shongum Pond, all in Morris County.

There are no natural lakes or ponds on the Whippany River, and Speedwell Lake on the Whippany River is shown on the map because it is a well known lake, but it is artificial. As regards the Passaic or Saddle Rivers, they have no natural lakes or ponds of sufficient size to warrant mention, and it seems superfluous almost to add that the beautiful body of water so well known as "Lake Dundee" is artificial, and simply impounds the water to furnish power to the factories at and near Passaic. By reference to the map it will be seen that a great number of creeks and brooks empty into each of the foregoing rivers.

With but few exceptions all of the lakes or ponds heretofore mentioned, are natural lakes. Greenwood Lake is thirteen feet above its natural level, having been raised to the latter height for uses of the Morris Canal. Before it was raised to its present height, however, and in its natural or primitive state, it was called "Long Pond." The rule seems to have been, to call all bodies of water, large or small, a "pond." In this connection, it may be proper to state at this time that Lake Marcia (a modern name), lying in the depression of the crest of the mountain at High Point, Sussex County, at an elevation of 1570 feet above tide, is the highest body of water in New Jersey. It is not on the Passaic watershed, however, and is noted here simply as an interesting fact. The highest elevation of the Passaic watershed is at the source of the Pequannock River, which is 1496 feet above tide. For location of lakes or ponds, with their elevation above tide water, see figures on map corresponding to figures in index.

Peckman's Brook rises near Verona, in Essex County, and empties into the Passaic, midway between Little Falls and Paterson; Oldham Creek rises in Bergen County, N. J., and empties into the Passaic, a short distance above the "Falls" at Paterson. Saddle River rises in Rockland County, N. Y., and traverses a great portion of Bergen County, N. J. It is fed by numerous small streams, the largest of which is Hohokus Creek, which unites with it at Paramus. Saddle River, as already mentioned, empties into the Passaic, on the Bergen County side of the river, at the town of Garfield, opposite the city of Passaic. Second and Third Rivers, two small yet important streams, also empty into the Passaic at and in the vicinity of Belleville. That portion of the Passaic which is above the inflow of the rivers already enumerated, is generally known as the Upper Passaic. This portion of the river, while smaller, and notwithstanding that it flows in great part through low and marshy meadow-land, is yet rich in the beauty of its scenery. Many of its windings through the valley are grand and picturesque. The meadows lying between Chatham and the mouth of the Rockaway River, as well as the Great Piece Meadows lying between the village of Pine Brook and the mouth of the Pompton River, at Two Bridges, are often entirely submerged. The sluggish flow of the water, clings tenaciously to the lowlands; and, whether it be tenable or not, it is claimed by some of the residents of the above named locality, that the water of the Pompton River, at its inflow into the Passaic at Two

Bridges, prevents the free flow of the Passaic; and in times of freshets, the water of the Pompton River forces that of the Passaic up stream, and thus causes the inundation of the Great Meadows. Expert engineers, however, have adjudged the inundation to be caused by a reef of earth and boulders at Two Bridges, and other points lower down the stream, at which points operations are now in progress for their removal, under the direction of State Commissioners. And it seems to the writer that this sketch would be incomplete, were it not stated that the late Prof. George H. Cook—so widely known and beloved in New Jersey, where he filled the office of State Geologist for more than a quarter century—was born on the banks of the Passaic, at Hanover. He labored long and earnestly to obtain relief for his native valley from the disastrous floods which the river he loved so well visited upon it in its capricious moods. He lived just long enough to see his efforts bear fruit; and the work of lowering the reef at Little Falls was only fairly begun, when he was called to cross that other river, the tides of which are so far beyond the control of mortal man.

Three huge rocks, or boulders, lie in the river at Two Bridges, and during the normal flow of the river they are exposed, and the view from the bridge, over the Passaic at this point, is at all times fascinating. The writer was told a legend in his boyhood days, wherein these same rocks were said to have been hurled by some giants from the high peak of the mountain at North Caldwell, at some Indian fishermen who were hostile to the giants; that the fishermen were crushed beneath the rocks, and that a certain species of fish were so frightened by the fall of the rocks, or boulders, that they left the waters of the Passaic, and never returned thereafter. Whether the legend be true, or not, we have no means of determining; yet, it does seem strange that a species of fish, different from any that now inhabit the waters of the Passaic, should have been found in *fossils*, at points so far distant as Boonton and Pompton; or that a portion of rock, identical with the three at Two Bridges, should still lie on top of the mountain at North Caldwell, five miles distant. These, and other evidences of a pre-historic age, gave rise, no doubt, to the legend, which is here given episodically.

The valley of the Passaic, whether taken in its entirety, or taken at random from almost any point, is fair to look upon; and many of the views are grand beyond description. Those from points near Millington, Stanley, Chatham, Livingston, Hanover, Swinefield and Pine Brook, near the mouth of the Rockaway River, are only equaled by the more extended views from Fairmount Avenue at Stanley, Riker Hill at Roseland, the school-house common at Caldwell, the Stony Pinnacle on Hook Mountain at Horse-neck, and the crest of Second Mountain at North Caldwell. The view from the highest point in Laurel Grove Cemetery, near Paterson, either looking up or down the valley, is worth a journey to see; and the same may be said of the views from East Side Park at Paterson. Garret Mountain and the Preakness Hills afford views ever pleasing to the eye, while the whole distance from the city of Newark to that of Paterson, is one grand panorama of enchanting landscapes, rarely equaled; and the writer has yet to see a river or valley affording views that surpass them. Many other fine views besides those above mentioned, lie along the banks of the Passaic; and the views contained within the limited pages of this book, give only glances here and there of New Jersey's famous river. Many small, yet beautiful tree-girt islands lie within the rims of its flowing tides, and offer restful retreat to any who may seek their cooling shades.

The Valley of the Rocks at Little Falls, as well as the Falls of the Passaic, at Paterson, are full of interest, and afford great pleasure to all who visit either of them. Aside from the rugged beauty of this scenery, the student of geology will find abundant study in the columnar trap-rock at Little Falls, or the basaltic rock at the Falls, at Paterson. And, while it is disappointing to many, and to be regretted by all, that during the dryer season of the year most, if not all, of the

water is diverted from the river to the race-way of the Society
for Establishing Useful Manufactures, yet the grandeur of the
open Chasm cannot fail but to impress, even with its barren-
ness of waters, the wondrous works of Nature's hand. And,
while it is admitted that the blasting of the rocks at Little
Falls, in connection with the operations now in progress for
the drainage of the Upper Meadows, has marred to some extent the natural beauty of this waterfall, yet the weird sur-
roundings of that charming spot are such that it will defy the
" march of progress" to blot out its many and irresistible fasci-
nations. While the windings of the flowing river, whether in
the upper, the middle or the lower valley, lose none of their
charm, but rather the charm increases the more one looks upon
it; and though age may creep upon everything around us, yet in
the flowing river there is a never-ending life; and its waters
run on to-day the same as when our eyes first looked upon
them years and years ago. Every murmur is just as full of
rythm; every ripple is just as bright, and sparkles with the same
brilliancy of beam; the moonbeams dance the same mystic
revels upon it; while the sunbeams fall with the same warmth,
and paint the same bright pictures. The autumn leaves fall on
its placid waters, and ride out to sea like little fairy ships, and
with each recurring season, the leaves will fall withersoever
the winds list, and ride on the bosom of the same river out to
the same sea. Tree-girt banks will cast their shadows just
the same, and the reflections will differ only as the trees shall
have grown larger, or given way to those of smaller growth.
And while the writer is free to admit that he has written wholly
from a sense of admiration for the river, and has striven to de-
scribe its many interesting features rather than give expression
to flights of fancy, yet he indulges in the hope that under the
guidance of poetic inspiration, some abler pen than his—

Will sing in rythmic line,
And all its way define,
And, in our hearts enshrine
The beauties of that river.

Paterson, N. J., 9th Mo., 1890.

www.ingramcontent.com/pod-product-compliance
Lightning Source LLC
Chambersburg PA
CBHW021530090426
42739CB00007B/862